I0464371

Learn Biology NOW:

Biology for the Person Who Has Never Understood Science!

By Justin Ascott

NOW Books

An Imprint of Minute Help Press

www.minutehelpguides.com

© 2011. All Rights Reserved.

Table of Contents

Introduction

Biology is an enticing subject. Knowing about how life began on the planet and how it operates is enormous fun. Biology is the study of living beings, such as microorganisms, plants and animals. It covers their origin, function, geographic distribution, interrelation with other organisms and reproduction.

Biology is organized into a number of disciplines and sub-disciplines, covering all the aspects of biological life. This gradation starts from the basic unit of life, the cell, which is the smallest functional unit of life. A living being can be a cell, tissue or an organism; a group of cells make up a tissue, a group of tissues make up an organ and a group of organs make up an organism. Understanding these concepts makes biology exciting, because, as a living being, you can relate it to yourself!

The Scientific Method

Observing living beings and musing over how they work is a typical day-to-day activity for curious people. Some of them conduct extensive experiments on these issues, while others read what has already been written on the subject. Scientists belong to the former class, and are consistently questioning nature. Comprehending how everything started and the biological process involved in the evolution, growth and development of living organisms is not a single step process. Analysis of the details of organisms requires a lot of observation and dedication from scientists. It encompasses an entire process of observing facts, setting hypothetical principles, conducting experiments, deducing and finally declaring the results. This is the basis for all classifications, functional analysis and medical sciences.

There are two ways of looking at any issue: One is the scientific, and the other is the unscientific, or the natural way. If you look at a green plant, the natural way understands that the green is the coloration of a pigment in the leaf. But for a scientist this is beyond what the naked eye sees. His perspective of anything in nature is to look for a reason for its presence. A series of hypothesis and experiments lead to the understanding that the green pigment is essential for the plant to grow, as it is the basic component for photosynthesis, a process by which plants make their food. These conclusions are laid down after a series of stages that can be summarized as below:

> 1) Observation: This stage starts with observation of the surroundings that can either be from a direct or an indirect source. A direct source could be self-questioning, such as how plants survive when their green color disappears in winter. Either way it is the first step of the quest that develops the chain of experimentation. This questioning of the observed facts leads to speculations from which the hypothesis is laid down.

> 2) Hypothesis: When a question is posed, an educational guess is laid down as a hypothesis. All further experimentation and declarations are based on this guess. For a scientist, there are certain conditions that determine the integrity of the hypothesis, like:

a) In every situation two cases A and B are taken into consideration: The hypothesis must be laid down under such constraints that if the experiment proves A to be false then it has to be B.

b) Secondly, the hypothesis should be one that can be tested. Many ideas can only be imagined and cannot be tested; these do not fall under the criteria of scientific testing.

c) Another salient feature is that the hypothesis must be revisited. Once proved or under process, any contemporary scientist should find this tentative explanation viable to further experimentation.

3) Experiment stage: This is the most crucial stage of the entire procedure, because it will either prove or disapprove the hypothesis. It determines the validity of the method being used and how it will be established to the world. The controlled method of experimentation is the procedure that has been the most widely adopted. In some cases where controlled tests are not possible, field tests are performed that are not always accurate, and cannot be established on a firm basis. The controlled procedure is conducted in different stages, which can be laid down as the following:

a) The essence of any experiment is to prove the hypothesis. During this process, the most crucial step is the collection of data. Limited samples are not considered viable experimental sources, as they are deemed to be biased. This makes collecting large samples from widespread areas vital.

b) Next is deciding the procedure of the experiment. Care in selecting the method increases the precision of the results, limiting hassles in later stages.

c) When the test is designed the factors must be under controlled conditions. This means that the duplicate sample should contradict the original. If one is false, the other should be true. Next is the confounding factor. These are factors that are not under consideration but might drastically affect the experiment.

d) Documentation of the entire procedure is necessary for future research. This means that when a hypothesis is proven either wrong or right and the results are published, it should be done so with clear details of the procedure so that when it is conducted by any other source it produces identical results.

4) When the entire process is completed other scientists scrutinize the documents. The experiment is then retested, and the hypothesis is declared either correct or incorrect. If it is proven wrong then another hypothesis is generated, which goes through the entire procedure again.

5) The final results are then published in scientific journals and are circulated. A number of scientists throughout the world can perform the experiment in an attempt to understand the facts. After several successful attempts, the hypothesis becomes a theory, such as cell theory, the theory of evolution and many more.

All this might sound appealing, but there are certain limitations to the scientific method. First, the methods are exclusively based on existing knowledge, as the experiments limit themselves to current instruments and theories. Second, a hypothesis is discarded if it cannot be proved. Third, these methods are limited to present discoveries. They do not answer any moral questions or the purpose of existence.

In spite of these limitations, it can be understood from further study on organisms that these scientific methods have formed the basis of much of what we know about the living world.

Fields of Biology

Everyday thousands of living forms are discovered that are collectively grouped, easing the complications of studying the subjects as a whole. They are then categorized into eight key fields, which include:

1) Botany: This is the study of the plants in the environment. It includes evolution, structure, reproduction, growth, metabolism, development and diseases in plants. The discipline is further subcategorized; for example the study of fungi is called mycology, the study of crops is called agriculture; horticulture is related to flowering plants, and so on.

2) Zoology: While botany broadly pertains to plants, zoology encompasses the entire animal kingdom. It relates to the study of the structure, embryology, evolution, development, growth, reproduction, metabolic activities and geographical distribution of animals. Sub disciplines in this field include ichthyology (the study of fish), mammalogy (the study of mammals), herpetology (the study of snakes), ornithology (the study of birds), and so forth.

3) Taxonomy: This branch of biology deals with the grading and naming of organisms. The classification is based on their cell structure, evolution and genetic models.

4) Morphology: This subject encompasses the entire forms and structures of the so-far recognized living beings. While zoology and botany include this study, the more specific group is morphology.

5) Genetics: This subject deals with genes and heredity mechanisms. The discipline relates to evolutionary principles and how traits are transferred from one generation to another, and includes cell study or cytology.

6) Ecology: The interrelation and acclimatization of the organism with its surroundings is studied in ecology.

7) Biochemistry: This deals with the energy transfer in and out of organisms and the chemical reactions that are required for existence.

8) Pharmacology: The essence of most of the discoveries relating to cells and organisms is related to this field, where diagnosis, treatment and prevention of diseases are studied.

There are many other fields that are placed in the following order to make their study easier. All of these are interlinked, and the study of one involves knowledge of other disciplines.

Cellular Biology

The cell plays a remarkable role in biology, as it is the functional unit of the entire living world. A unique feature of the cell is that it can exist both as an individual organism like those of bacteria or form aggregates and organize into tissues and organisms like in humans. It is astounding to know that humans are a storehouse of more than 100 trillion cells.

In the human body, the smallest cells are about 4 micrometers, and the largest are those found in the anterior horn of the spinal cord. The largest cell in the world is the unfertilized ostrich egg, which is about 3.3 pounds. The study of such facts and their interrelation with the environment is called cytology, or cellular biology. It includes their physiological properties, organelles, their function, life cycle and division.

Techniques to Study Cells

The concept of the cell first came to light when Robert Hooke observed them in a piece of cork under a microscope. After this there was a constant evolution in the principles and theories as new and sophisticated technologies were invented.

The traditional microscope was confined only to the framework of the cells. When the optical microscope, transmission microscope, scanning electron, Fluorescence and Confocal microscopy and other techniques were introduced, it detailed the organelles and their function.

Cells are now isolated from their tissues and studied under high resolutions. Identification of genetic material such as DNA was another turnaround in the field of biology. The study of these was conducted through certain methods like spectroscopic quantification, DNA micro array, gel electrophoresis, chemical sequencing and many more. With these developments, definitive scientific experiments were conducted leading to the postulation of cell theories.

Cell Theory

Starting from when cells were first observed by Robert Hooke, there have been a number of scientists who have enhanced this discovery. The idea of cells and their relation to the existence of life was then postulated as the early classical cell theory. It states that:

1) All living organisms are made up of one or more cells, and they form the basic unit of life.

2) Cells arise by division of the previous cells.

3) The cell is a unit of organization and physiology of living beings.

The modern theory has a few modifications which state that: the energy flow in the organisms occurs within the cells, cells contain the genetic information that is passed during cell division, similar species have similar chemical compositions of cell materials, the organism's activity depends on the cells, and organisms are either single celled or multi celled. Biologists have laid this theory down as the base for all further investigations.

Types of Cells

The classification of cells into two broad groups eases the complications that arise due to the various characteristics that the cells of individuals display. This is based on morphological and functional similarities. In general all the single-celled organisms that lack membrane-bound organelles are grouped under the prokaryote class, and those that have a membrane around their organelles are grouped under the eukaryotes class.

> 1) Prokaryotic cells: These cells are simpler, smaller and unicellular organisms. They are broadly classified as Bacteria and the Archaea. Apart from the fact that they are unicellular and simpler than the Eukaryotes there are other reasons for their grouping. The cytoplasm region of these cells is intriguing as it differs markedly from the complex structures. Typically these have organelles without membranes dispersed in their cytoplasm or cell material. The cell is covered with a plasma membrane and an outer cell membrane. Some have another outermost layer called the capsule for protection. The DNA is condensed to form nucleoids, circular chromosomes and ribosomes. They also contain an interesting antibiotic-resistant plasmid that is the distinctive feature of these cells. They even have flagella, pili or cilia around their cells.

2) Eukaryotic cell: These are the most interesting, as they include the animal and plant cells that are seen in the environment in day to day life. One of the major differences between the prokaryotic and eukaryotic cells is the presence of membrane -bound organelles. The genetic material is enveloped in a membrane that is collectively called the nucleus. The DNA is a linear molecule called the chromosomes with histone proteins. Other organelles are the mitochondria, endoplasmic reticulum, Golgi apparatus, vacuoles, lysosomes, centriole and vesicles. These are further divided into plant and the animal cells. Plant cells have a cell wall, larger vacuoles and lack centrioles.

The entire effort by scientists to understand the different cell organelles and their function is focused towards the movement of proteins and other substances in and out of the cell membrane. These are correlated to diseases, pharmacology, biochemistry, genetics and other branches of biology.

Collectively all the organelles of the cell participate in the different processes either directly or indirectly. One of the most fundamental processes is the protein transfer. Proteins are synthesized in the ribosomes by an enzyme-catalyzed process called biosynthesis. Other processes include active and passive transport of the molecules in/out of the cell, autophagy (a means by which cells ingest their own cell components or external invaders), adhesion to other cells and tissues, reproduction through cell division, cell movement, respiration, photosynthesis and gene expression.

The growth of cells is another process that is studied extensively from its inception. Binary fission, Mitosis and Meiosis are the three types of cell divisions that are considered asexual reproduction. In all the three processes, there is a division of the cell along with the chromosomal material. The growth process can be encountered in daily activities like the regeneration of cells on a damaged skin surface. The netted dark brown structures, which are observed when the skin is ruptured, are the dead cells. Regeneration of these cells is the healing process.

Reproduction and Development

The essence of any living form in the world is to propagate their identical offspring through a biologic process called reproduction. This is a phenomenon in plants, animals and microorganisms. The modes of reproduction are broadly categorized into sexual and asexual methods. The sexual method involves gametes from the two opposite sexes that follow a series of division and fusion processes, resulting in the offspring.

Asexual reproduction is limited to unicellular organisms, mostly plants and with the exception of a certain ant species in the animal kingdom. In this type, the individual can reproduce without the involvement of the opposite sex. But whether it is the asexual or the sexual, the primary process in reproduction involves the division of the genetic material and its propagation to the next generation.

The growth and development of organisms is an irreversible process that is a consequence of reproduction.

Asexual Reproduction

A) Binary fission: This is a common method of cell division in most prokaryote and a few Protista. This process results in the division of the cell and the nuclear material into two parts, which then have the potential to grow to their original sizes. The first stage in the process is the division of the DNA material. After this the cell membrane divides with its corresponding circular DNA, and finally the cell replicates into two different organisms.

B) Budding: This is most common in yeast cells. It can be observed on many substratums as the colored spots spread wider from one day to another. This reproduction is called budding. Some cells of the mother organism form small aggregates. These then isolate themselves from the mother on maturity, giving rise to a new organism.

C) Vegetative production: Observed in plants, this can also be induced manually. They are propagated by vegetative parts of the plant through treatment of hormones while some show natural growth. Potatoes, onions, strawberries, and grapes are all examples of natural vegetative propagation. Tubers and bulbs stemming propagation through stolon and suckers and leaves stemming propagation through the buds on the leaves (like in Bryophyllum) are all some examples. Manual intervention in vegetative reproduction is common by horticulturists, because such processes can be controlled by the addition of chemicals. Roses are another classic example for such treatments. Stem cuttings of the plant are planted to give rise to a potentially new flowering plant. Other types are layering, budding, grafting, division and tissue culture.

D) Spores: Most plants, Algae and fungi reproduce via the sporogenesis. This process involves the formation of spores in sacs which, when matured, burst to disperse the spores. When these reach the right substratum they develop the roots or grounding structures underneath and the vegetative parts above. This gives rise to a new offspring.

E) Fragmentation: A new organism grows and develops from a part of the parent. This is seen in many insects such as earthworms.

While these are the asexual means, sexual reproduction is the only mode of reproduction at the macro level in all humans and a number of animals. Plants reproduce through the fusion of a male and a female gamete. During sexual reproduction the number of chromosomes in each parent cell, the mother and the father cells, are divided into two halves.

The mother and the father chromosomes undergo fusion that restores the diploid nature. Then the development period starts giving rise to new offspring. The process of genetic inheritance is similar in both plants and animals, but there are slight differences regarding the division of the material:

A) Plants: The colorful flowers that are seen on plants are their reproductive organs. Male sperm is produced in the pollen grains and the female is produced in the carpel. The long strings with tiny heads in flowers are the stamens or the male organs and the tubular structures with bulb like end and sticky opening is the carpel or the female organ. When the pollen in the stamens is dispersed onto the carpel, the male cell develops a long tube called the pollen tube into the carpel. The sex cell nucleus is then transferred through this to the ovule to fertilize the egg cell. A process called double fertilization takes place where an egg cell and surrounding tissues in a fruit are generated. Now the new plant in the form of fruit is ready to develop.

B) Animals: Most animals reproduce through the sexual process, in which the male gametes fertilize the female gametes. But the process of fertilization and development largely varies in different phyla. Considering mammals, which include humans, the process of fertilization takes place internally within the female reproductive system. After copulation, the sperm travels through the vagina and fertilize internally in the uterus of the female. In the following series of divisions a group of cells implant to the walls of the uterus. This is the beginning of the embryogenesis. Once it is well developed the fetus is propelled out through contractions of the uterus. The entire cycle of the development into an individual is called the gestation period, which varies in different organisms. A few animals such as birds and reptiles develop the eggs outside of their bodies. All of these are macro level reproductive processes.

The micro level takes place by cell division. Cells divide and proliferate by the process of mitosis and meiosis. Single celled organisms follow the binary fission method. Mitosis takes place in the somatic or the body cells. This is the regeneration process that maintains the cell count in the body when the older ones die. Meiosis occurs in the germ cell or the sexual cells.

The paternal and maternal chromosomes from the sperm and egg undergo a series of steps to provide the genetic makeup that results in the traits of the offspring. In both processes, the cell material and organelles are generated.

Genetics

The crux of living existence lies round the DNA and RNA, which are considered to be the genetic material in the cells. Some organisms have both as their genetic material while others use one of them.

A branch of biology that details topics such as gene inheritance and variations in living organisms is Genetics. It involves the study of molecular structure and the function of genes to understand the inheritance of certain traits from parents to their offspring. Modern understanding of the implications of Genetics came into prominence starting in the mid 19th century.

Genes are a part of DNA that in turn is made up of 4 different types of Nucleotides, which are molecules that play a central role in metabolism. Genetics play an immense role in appearance and behavior of an organism, although they may alter to a varied degree based on the experience of the organism after its inception.

Theories revolving around genetics have undergone many changes, where a theory that once was assumed right has been proven wrong later. Some examples of such instances are the theories that evolved before the 19th century, which perceived that traits are inherited as a smooth mixture of traits from both parents.

It was proven later that inherited traits are a combination of varied genes. Genes exist on chromosomes that constitute proteins and DNA. The DNA has a helical structure and is the one that transmits inheritance. Nucleotides are pointed structures, which form a ladder-like a skeleton in the helical model.

The DNA controls the protein production, where the template is used for creating a matching messenger RNA. These RNA generate the amino acids sequence and this translation between nucleotide and amino acid sequence in known as the genetic code. When this was discovered the isolation of specific DNA fragments and their studies became possible.

Mendel's Laws

Inheritance in organisms occurs by means of traits called the Genes. This was first discovered by Gregor Mendel, and the laws laid by him are termed Mendel's laws. His scientific experiments were conducted on the pea plant. The pea, being a diploid, consists of genes with two discrete versions called the alleles. Each one is inherited from each parent. For example, if the mother is denoted by Aa and the father by Bb. A and a are two alleles, the same with Bb. After fertilization, the probability of alleles in the offspring is four: either of AB, Ab, BA, and Ba. The traits inherited by the offspring are highly dependent on the gene inherited. This is Mendel's first law or law of segregation.

Chemical Base for Inheritance

Chemically DNA is Deoxyribonucleic acid, which is a chain composed of four different nucleotides: Adenine, Guanine, Cytosine and Thymine. The genetic information of any living being lies in the way these four nucleotides are bonded. When the DNA is observed it looks like a double helical strand with the nucleotides pairing with its partner on the opposite strand. Adenosine pairs with Thymine and Guanine with Cytosine to form rings.

The entire strand along with proteins is organized into chromatin that, with some histone proteins, collectively is called the Genome. Living beings are either haploid or diploid. Haploid organisms have one set of chromatids while the diploid come with two sets of chromatids. In the diploid two alleles are inherited from each parent.

Why the Gene Plays an Important Role

Genes express their functional effect through the production of proteins. Long chains of amino acids, which are responsible for the activities of the cell, are transcribed from the nucleotides present in the DNA strands. The strands get transcribed into a specific RNA molecule that acts as an intermediate in the entire process. Three consecutive pairs of nucleotides on the RNA strand are considered a codon. Each set can be translated to one of the twenty amino acids.

The trio sequence can also truncate the process of protein formation. Thus, the chain flows from DNA to the RNA and to the amino acids. The resulted proteins facilitate in a number of functions like hemoglobin, enzymes, and carrier molecules in and out of the cell. Each chromosome contains thousands of genes, but not all genes are expressed at the same time. Certain transcription factors are regulatory proteins, which initiate or terminate the process of protein formation.

Reproduction

It is known that cells divide, and this is the reproduction at the cellular level. But when cells divide the primary exchange that takes place is the gene that forms the basis of inheritance in organisms. When dividing asexually there are identical genes that are transferred from the parent to the daughter and these are called clones. But in most of the eukaryotes the genetic material is inherited from both parents.

A series of alternates between haploid and diploid sets takes place resulting in the recombination and linkage of the chromosomes creating new shuffled stretches of DNA.

Drift in Nucleotide Sequences

Any shift in the nucleotide can drastically change the activity of the cell, leading to diseases, improper functioning and other impairments. In some cases during the replication of DNA the second strand produced occasionally exhibits an error in the nucleotide sequence.

This is called mutations and impacts the organism. Substances that trigger this process are called mutagenic, which can either change the base pairing or damage the DNA structure. Damage caused by UV rays to the DNA strands results in a complete change in the cell functioning. The strands try to regain their original pairs, but in few cases this goes wrong and the error remains permanent, creating faulty protein coding.

Genotypes are the gene expressions, and the characteristics they express to the human eye are the phenotypes. It is these phenotypes that are changed when the genotype changes. Evolution of organisms is related to this drift in the genotype.

Evolution and Heredity

It is often heard that man is descended from ape. But what is this change, and why has this occurred? This is answered by the evolution of organisms through generations. In this process, they inherit their traits from their parents, forming the basis for heredity.

These include anatomical, physical and behavioral patterns of the organism. Percolation of genes from one generation to another occurs through the process of reproduction. Though heredity is considered for each generation, evolution is not a single generation process.

This occurs through ages, which results in a variation in the population. Gene expression, mutations, and gene flow are the various reasons for the diversification of traits. Ultimately it is crucial to identify evolution and heredity as consequences of Genes.

Evolution

The phenotype of any individual depends on the combined effect of the genotype and external factors. If a person from a tropical area has inherited the trait of dark skin coloration from their parents, the phenotype does not remain the same throughout. When the climatic conditions change or geographical location changes to the colder regions a slight change in the original coloration occurs.

This is a combined effect of climate with gene expression. Natural selection is the phenomenon of life on earth, so the evolution of a genotype depends largely on how strong the new allele supports reproduction and fitness in the society. This can happen by various means, which are intentional and natural. Sources of evolution can be categorized as:

A) Variation: There are slight genetic variations when the external environment affects gene expression. Through a due course of time, this becomes the dominant allele. Now when reproduction occurs this allele is transferred to the offspring, bringing out a permanent change in the expressed trait. Thus, evolution has occurred.

B) Mutation: This is another cause for evolution. There can be spontaneous mutations or induced mutations. These processes can create both harmful and beneficial properties. Some result in the duplication of the chromosomes increasing the genes in the genome. In some cases, there are entirely new traits that are added. Such are percolated to the next generation.

C) Gene flow: When genes between different populations and species undergo the reproduction process, the resultant genome consists of alleles that demonstrate entirely different traits. This is gene flow. Such conditions can result in offspring that are totally different from the society. When survival becomes competitive in these scenarios further mating is prevented. Some retain their compatibility and proceed with the evolution process. A classical example for gene flow is the mule, which is a cross between a donkey and a horse.

There can be genetic drift if the adaptation of the species through these processes becomes sustainable, but the time for fixed drift depends wholly on the population. Such evolutionary adaptations increase the fitness of the organisms, making them more comfortable in challenging environments. Conditions may arise when the micro level of evolution, which starts in the genes, becomes an environmental factor in due course of time.

This leads to a new factor called the speciation, where a group from the society departs and a new group incorporates itself giving rise to a totally different pool of genes. When a cross in this new society takes place, alleles and traits change. Thus, this process becomes constant. A few traits become entirely extinct and eradicate all the corresponding species. Dinosaurs are an example for this.

Heredity

The passing of traits from the parents or ancestors to the offspring is heredity. Scientifically it is the transfer of genes to the next generation. Common observation of hair coloration, eye colors, and behavior is all related to the parents or grandparents. These phenotypic expressions are controlled by the inherited traits. There are two types of heredity: dominant and recessive. Dominance is the character of the allele that is expressed in every generation.

The recessive trait is the other less dominant or nearly zero trait. The green pea experiment is the most famous example. These have two genes, one for the yellow pods and the other for the green. It is a common observation that peas appear green and exceptionally rare conditions provide the yellow coloration. If the green color is denoted by G and the yellow by g, most of them inherit the green. But when the alleles in the chromosome are gg then a yellow colored pod is created.

Heredity is categorized by the number of loci in the DNA or the chromosomes involved. It is also influenced by environmental reactions, sex linked inheritance through male or the female, or over-dominance or gene coupling.

The evolution of organisms is comprehended by the study of fossil fuels. These are remnants of organisms that are found on excavations of the earth's layers. It is considered to be a gradual change of hereditary traits through the ages and is a continual process. These principles are laid down by the observations of scientists and understanding of the genome activities.

The theory of evolution was proposed by Charles Darwin in 1859, which only mentioned that traits were transformed from one individual to another. In 1865 the Mendelian theory was proposed which was later established as the framework for the current modern evolutionary synthesis. Apart from being a boon to the development of organisms, evolution and heredity principles are widely correlated to other fields.

Their key usage is in the diagnosis of genetic disorders and their treatment, with the artificial syntheses of enzymes and antibiotics utilizing these principles. While evolution has increased the adaptation of life on earth, it has also brought about extinction of thousands of species. This is a continuous process that can terminate many current life forms. These derivations have made the classification of organisms utilize the principles of heredity, evolution and genetics.

Classification

There are millions of plants and animals on earth, and many of these have become extinct. Identifying all of them and studying their characteristics is beyond human scope. To make this simpler, scientists categorize them into groups based on similarity in characteristics, or more precisely their traits. Classification can be more easily interpreted as a family tree.

The hierarchical classification is called taxonomical rank in biology. There are seven major ranks that have been accepted by most biologists: kingdom, phylum, class, order, family, genus and species, among which species is the smallest. Species is defined as the identity of the individual. It is like addressing an individual with his proper name rather than the generalized common name "human".

The earliest classification started when Aristotle grouped animals based on their reproductive systems. Following this, many eminent scientists classified living beings based on the knowledge they acquired during their lifetime. The most significant milestone in the field started in the 17th century when John Ray described in his book Historia Plantarum the taxonomy of plants based on observed characteristics.

The next step on this journey was by Carolus Linnaeus in his Systema Naturae, where he used five ranks: class, order, genus, species and variety. Plants and animals were grouped into two kingdoms: Animale and Vegetabile. With the advent of the microscope single celled organisms came to light. The classification was again given a new direction with the addition of a new kingdom, Protista.

The third kingdom Protista was then further divided based on the presence and absence of nucleus into Monera and Protista, where the former included those without a nuclear membrane.

When Fungi were discovered, they displayed characteristics that included plants, animals and the prokaryotic features. This led to the addition of another kingdom: Fungi, upon which the grade was coined "the famous five kingdom classifications", with Monera, Protista, Plantae, Fungi, and Animalia. All these support the Darwinian principle of common descent.

With the gene becoming a dominant source for all evolutionary bases during the late 1960's Cavalier-Smith's six-kingdom classification was proposed. At this point, the traditional ranks were replaced by evolutionary trees. The new phyloclads were intended to exist along with the primitive rank system. This trend is called the cladistic taxonomy, which divided organisms as:

1) Empire Prokaryota: Kingdom Bacteria

2) Empire Eukaryota: Kingdom Protozoa, Kingdom Chromista, Kingdom Plantae, Kingdom Fungi, and Kingdom Animalia.

The classification tree consists of these principal divisions: Life-Domain-Kingdom-phyla-class-Order-family-Genus-Species. These are further grouped into sub and super categories such as sub family, super family, sub phyla, super phyla and so on. Classification is changed constantly as new species are identified.

Though the grouping system has been under a continuous process of change, the usage of genus and species is still followed for the universal identification of organisms. To avoid the confusion of different languages, a Binomial nomenclature was introduced by Carolus Linnaeus, in which all names are given Latin versions. They are named according to their genus and species names. Humans are named Homo sapiens where the first is the genus classification.

Prokaryotes and Viruses

Prokaryotes and Viruses are all predominantly unicellular organisms with a few exceptions like myxobacteria, which have multi cell stages in its life cycle. Most of these are infectious and harmful to the living world. They are invisible to the naked eye and are studied under microscopes.

One key reason for the classification of prokaryotes into a separate taxonomical group is the absence of membrane- bound organelles. All the organisms are classified into two groups, namely bacteria and archaea. They acclimatize to every habitat, from extreme temperatures to the normal, human sustainable climates.

These are singular but live in closely associated colonies and can be encountered as thin layers called the Biofilms, and are seen attached to solid surfaces, air-liquid and liquid-liquid interfaces. Another interesting fact is the cell to cell signaling within the film. The aggregation displays a collective and social behavior with each other. These are found in four different shapes: circular, rod shaped, spiral or comma shaped.

Their cells contain an external flagellum, which is used for locomotion. The nutrition types can be through sunlight, organic or inorganic compounds. Some bacteria lead a symbiotic relationship with humans like those that reside in the human gut and assist in digestion.

In the rhizosphere zone they adhere to the roots of leguminous plants and take part in nitrogen fixation from the soil to the plant, and in return they obtain processed food. Besides these many of them are parasitic, causing infectious diseases. Some are even fatal, and many have not reached the level of diagnosis or treatment.

Viruses are the most interesting among all organisms. This pertains to their morphology and the startling rates at which they duplicate, resulting in fatal diseases. They share a structure that is unique from all living cells. A single virion consists of a protein coat called a capsid and the shape of the structure can be helical, icosahedra, envelope or complex. Their only feature that matches with other cells is the genome.

They can have RNA or DNA or both as their genetic material. The strands can be single, double or both with different shapes. They multiply at rapid rates where the genetic material is injected into the host cell. It is then replicated after which the protein cases are developed. As a result, a new virus is generated. Some incorporate their genetic material with the host and reproduce along with the host cell.

Viruses infect all organisms, including prokaryotes, plants and animals. They spread in many ways from one organism to another; for example, in plants it is spread by insects and in animals by blood sucking insects. Apart from these they also spread via sneeze, cough, food or water and so on.

Organisms develop an immune response to these viral infections but sometimes these viruses evade the immune response. There are also some Bacteriophages that are harmless to plants and animals, but a few are responsible for the most harmful diseases like AIDS and cancer. Research is still ongoing to treat these diseases, though many antiviral drugs have been discovered.

Protista and Fungi

Protista and Fungi are part of a large group of eukaryotes that are organisms with cells that contain complex structures. They are classified as two different kingdoms and display diverse characters.

Protista is a diverse group of 30 to 40 types with each being either unicellular or multi cellular without specialized tissues. They are distinguished from fungi, plants and animals with regard to the basic cell structure, life cycles and motility mechanisms.

They survive in any environment with water. There are some such as kinetoplastids and apicomplexa that are responsible for diseases such as sleeping sickness and malaria, while others like green algae are photosynthetic. Thus, they are either photosynthetic or organic compound feeders. Their mechanism of feeding involves engulfing the feed through phagocytosis, where the cell material along with the membranes circles the feed.

After this, it is digested by the cell. Reproduction in these organisms is both sexual and asexual. Asexual reproduction is done through binary fission and the sexual mode is through the formation of gametes. Some have complex life cycles like the malaria parasite. It undergoes varied stages in the red blood cells of humans leading to chronic effects.

Some of the commonly known fungi are yeast, molds and mushrooms. These are classified into a separate kingdom along with plants, animals and Protista. The kingdom is divided into one subkingdom, seven phyla and ten subphyla. They cover a wide variety of unicellular and multicellular organisms, which live in diverse habitats including soil and dead matter.

A few species are found as symbionts with plants, animals and other fungi. These are specifically of interest to many scientists owing to the characteristics they display. They are similar to animals in the lack of chloroplasts and a need for a chemical substratum as an energy source. When compared to animals they have similarities in the presence of a cell wall and vacuoles. Besides these, they show resemblance to bacteria and Euglenoids. Unique to Fungi cells is the cell wall with chitin, which is a derivative of glucose and a long-chain polymer.

They grow as elongated filaments spreading far and wide. These reproduce both sexually and asexually. Asexual reproduction is through conidia or fragmentation. Sexual reproduction shows immense diversity and differs from plants, animals and within the kingdom.

Both Protista and Fungi have organisms that are parasitic and useful. In the former certain species like the algae are potent sources of the ecosystem. Spores from a few species are utilized for controlling other plant diseases. In the kingdom Fungi, many species display a mutual relationship with other beings called symbiosis. They express these relations with plants, algae, cyanobacteria and insects. Apart from these some cause diseases in plants and animals like the rice blasts, ringworm and athlete's foot. When left untreated some are fatal to life. Some mushrooms are edible; a few are used for medicines, food culture, and pest control. The study of this is called mycology, which is considered to be a part of botany.

Plants

Living organisms that belong to kingdom Plantae are known as plants. Some of the familiar organisms of this clan are trees, flowers, herbs, bushes, grass, vines and so on. The branch that deals with their studies is called Botany. There are around 315,000 identified species of trees and 281,821 species of flowers. Most of the plants obtain their energy from sunlight through a process called photosynthesis, which uses chlorophyll in chloroplasts.

Plants can be broadly classified as green algae, tracheophytes, bryophytes and seed plants. The former are vascular land plants, and the latter are non vascular plants.

Green Algae

Green algae can be further divided into chlorophyta and charophyta. There are around 380 living species of chlorophyta and about 4000 to 6000 species of charophyta. Most of the green algae generally live as single cells, with some of them forming colonies. A few organisms maintain a symbiotic relation with green algae for obtaining energy through photosynthesis from them. They have bright colors owing to their possession of chlorophyll a and b.

Bryophytes

Bryophytes are non vascular plants that generally refer to all embryophytes, or land plants. Though some of them have tissues that carry water, they do not contain lignin and so are not categorized as vascular tissues. Examples of bryophytes are liverworts, mosses and hornworts. Liverworts are small in the range of 2 to 20 mm wide and around 10 cm long.

There are 10,000 identified species of liverworts. Mosses are small and soft with a length of 1-10cm, with a few exceptions that can be much larger. They generally grow close to one another in damp and shady locations. Hornworts are non vascular plants with an elongated horn like structure called sporophyte. These, like other bryophytes, are generally found in places damp and humid. Some of them grow as tiny weeds on garden soil or cultivated lands and on the bark of trees. There are only around 100 identified species of hornworts.

Pteridophytes

These are vascular plants that do not produce flowers or seeds; hence they are also called vascular cryptogams. Pteridophytes are plants with xylem and phloem. Xylem and phloem are transport tissues, with xylem used to transport water and phloem utilized for the transport of sugar or glucose. Xylem is used sometimes to transport nutrients. The sexuality of pteridophytes can be classified as dioicous and monoicous.

Dioicous pteridophytes produce either male organs or female organs, only on a single gametophyte body. Monoicous pteridophytes are ones that produce both male and female organs in the same gametophyte body. These can be further distinguished based on what they produce first; if they produce male first followed by female then it is known as protandrous pteridophytes. If they produce female organs and then male they are called protogynous pteridophytes. Club mosses, spike mosses, and ferns are all examples. Ferns are used in medicines, weeds and grown as food.

Seed Plants

Plants that produce seeds are seed plants. They are also called spermatophytes or phanerogams. They can be divided into five groups, namely cycads, ginkgo, conifers, gnetophytes and angiosperms. Seed plants can be broadly classified as gymnosperms and angiosperms. Gymnosperms are regarded as the natural group and cycads, ginkgo, conifers and gnetophytes are different species of gymnosperms.

Gymnosperms

The term gymnosperms in Greek mean naked seeds. Conifers are the largest group of living gymnosperms, followed by cycads, gnetophytes and ginkgo. There are about 700 to 900 distinct species of gymnosperms, with 600 to 630 of them being conifers, 130 cycads, 75 to 80 gnetophytes and 1 species of ginkgo.

Conifers are cone bearing seed plants with vascular tissues, most of them are trees with some being shrubs. They are of high economic significance primarily for paper and timber production. Seed plants with heavy, woody trunks and leaves are cycads. The trunk size of cycads can vary from a few centimeters to a few meters length and are generally found in tropical and subtropical parts of the world.

Angiosperms

Angiosperms are seed producing plants like gymnosperms, but they differ from gymnosperms with the characteristics like flowers, endosperm and fruit production. The major difference between the two is that in this category the seed is enclosed within a fruit. The complexity of the vascular tissue reaches summits in these flowering plants. They are broadly divided into monocotyledon and dicotyledonous plants. Cotyledons pertain to the character of the seed. As an example rice is considered monocotyledon and peas are considered dicotyledons. These plants experience complete life cycles with stages including flowers, fruits and seeds.

The cell structure of the plants differs significantly from other organisms. The cell is covered with a cell wall and contains large prominent vacuoles and pigment carriers called plastids. The cells lack centrioles and have a few mitochondria. Another distinguishing feature of most plants from other kingdoms lies in the energy production through the process called Photosynthesis. They contain a pigment called chlorophyll that is used in the process. In this way they take an active part in the ecological balance of Earth.

Certain algae are a boon to the marine ecosystems. Some like the Venus flytrap are carnivorous, where they trap flies between their spiked leaf margins and kill them. Energy is then derived from the nutrients of the fly. Plants display many features like a response to stimuli, immune systems and movement. Locomotion is absent in plants and there is an internal vascular system that transports water and nutrients. Reproduction in plants includes both sexual and asexual reproduction.

The economic importance of plants is obvious. They are used in agronomy, horticulture and forestry. As food they form a staple diet for many animals. Wood and other cell saps and secretions from plants like gums and resins are utilized for various industrial purposes.

Flowering plants and many ferns have ornamental and aesthetic uses. A few are harmful, such as weeds and hay fever due to flower pollen. Though differing greatly from animals and leading a sedentary life there are plants which are said to live for hundreds of years, forming a reliable source for scientific research.

Animals

Kingdom Animalia is the largest of all of the six and has an enormous diversity in organisms. One common feature in the entire kingdom is the multicellular eukaryotic cell organization. Most animals can move independently and spontaneously in the process consuming energy. Animals being heterotrophic are different from plants and algae as they use organic carbon for growth. They also differ from fungi because they do not possess rigid cell walls.

Except for sponges and the placazoans, all animals have distinguished bodies that are differentiated into tissues. The organization of the body into systems, like the muscular, digestive, circulatory and nervous system makes them classified as Eumetazoans. Stimulus by these animals is a collective result of all these systems.

Animals are classified broadly based on the presence of vertebrae into vertebrate and invertebrate:

1) Invertebrates: These are animals that lack a backbone or the vertebrae. Accounting for more than 97% of the animal kingdom, these include all those that are not grouped under the subphylum vertebrate. Common examples included in the subphyla are sponges, flat worms, sea urchins, star fish, all insects, earthworms and so on.

2) Vertebrate: This subphylum includes all chordates with about 58,000 species. The sub-phylum is divided into classes agnatha, chondrichythis, and osteichthyes, Amphibia, Reptilia, Aves and Mammalia. These are the jawless fishes, cartilaginous fishes, bony fishes, amphibians, reptiles, birds and mammals in the corresponding order. Common examples of these are fish, snakes, lizards and man.

While this is the classification based on the sub-phylum, they are organized into subkingdoms that rank them according to their development and their symmetry. There are two subkingdoms: Parazoa and Eumetazoa:

1) Subkingdom Parazoa: The only existing animals from this division are the sponges from the phyla Porifera and one from the phyla Placozoa. They are asymmetrical, and their body is not differentiated into tissues and organs. This makes them unique.

2) Subkingdom Eumetazoa: All of these animals have their body organized into tissues and show gradual evolution through the phyla to form complex organs. Gradation of the subkingdom is done based on the number of symmetrical halves the body can be divided into and the formation of the blastomere. They are placed under two categories: Radiata and the Bilateria.

A) Radiata: These animals display radial symmetry and they contain only two germ layers: endoderm and ectoderm. It consists of the jellyfish and coral reefs.

B) Bilateria: All animals with bilateral symmetry are included, and these developed a front and back opening during the embryonic stage. That is, humans and animals with a mouth and anus as the front and back apertures are categorized under this group. Another feature is that bilaterians develop from three germ layers: the endoderm, mesoderm and the ectoderm. With the exception of flat worms and round worms, all others have their body cavities filled with ceolom. These are further classified as:

a) Deuterostomes: The major difference of these lies in the formation of the anus as the first opening of the digestive tract during the embryonic development. The mouth appears secondary to the anus. The embryonic cleavage varies significantly from other animals. Echinodermates and the chordates are the main phyla that include all the vertebrates and some like the sea urchins, star fishes and sea cucumbers from the invertebrates.

b) Protostomia: Though not ranked, it includes all those that develop a mouth as their first opening. Superphyla Ecdysozoa, Platyzoa, and Lophotrochozoa, are all included under this category.

1) Ecdysozoa includes those that undergo molting during their life cycle. Insects, spiders, crabs, and roundworms are all classified under this superphylum: Of these, the largest phylum is the Arthropod. The ceolom present in the body cavity is called pseudoceolom.

2) Platyzoa consists of flatworms or the Phylum Platehelmenthis. Most of these are parasites like the tapeworms that infect the digestive tract of humans, Few other phyla are also included, but they are all predominantly pseudocoelomates and microscopic.

3) Lophotrochozoa includes the mollusks and the annelids. These include snails, earthworms, octopods and others.

All animals feed directly or indirectly on other living things and can be classified as carnivores, herbivores, omnivores and parasites. Most animals feed indirectly on sunlight. This happens as they feed on plants or other organisms, utilizing the photosynthetic product, or sugar through the process of glycolysis. There are some exceptions though, such as animals living on the ocean floor that are not dependent on sunlight

Carnivores

Carnivores are meat-consuming organisms that derive their energy and nutrient requirements from animal tissues. They can be classified as obligate and facultative carnivores. Obligate carnivores are those that derive that energy from animal flesh while facultative are those that derive their nutrient requirements from non-animal food. Carnivores can be further classified as insectivores when their main source of food consumption is either insects or invertebrates, and piscivores, when they eat only fish.

Herbivores

Herbivores are those organisms that primarily consume plant life. Herbivores consume plants, algae and photosynthesizing bacteria. They form a vital link in the food chain as carnivores feed on some herbivores, thereby gaining the carbohydrates they receive via consumption of plants.

Omnivores

Omnivores are those animals that eat both plants and animals as their primary source of food. Some particularly common examples of omnivores are humans, pigs and cows.

Parasites

Parasites are organisms that benefit at the expense of other organisms; the other organisms are called hosts. They are referred to as organisms with life stages beyond one host. Examples of these are vertebrate hosts and tapeworms. Parasites are smaller than their hosts and reproduce at an alarming rate compared to their hosts. Parasites can be classified as ectoparasites and endoparasites based on where they live. Ectoparasites live on the surface of the host, whereas endoparasites live inside the body of the host.

Reproduction in animals includes both sexual and asexual, but more often it is the latter. From the poriferans to the humans there is a significant development in the organization of the cells and the body organs.

Ecosystems

Living beings are related to the living and non-living world around them. They colonize with other similar species, intervene with activities of nature, and obtain energy from other organisms or by utilizing environmental factors. This entire cycle of interrelation leads to the formation of a potential community. An ecosystem is the interrelation of the abiotic and biotic factors of a region.

In an ecosystem the most fundamental fact is a continuous flow of energy and nutrients. Food produced by certain living beings is converted to energy that flows to the consumers once they die, or through their excreta to the secondary consumers, which result in the purification of the toxic substances. This is called the food chain. A classic example of this is the following: Plants manufacture food by means of photosynthesis, energy is transferred to animals which are the primary consumers, and a number of microbial feed on their excreta.

Communities can be temporary or permanent. Ecosystems are classified as terrestrial and aquatic. Divisions include the regions in which the community survives like: fresh water, tundra, savannah, desert, coral reefs, terrestrial, forest, pond, marine and so on.

Biomes and Biodiversity

Widespread ecosystems with common traits can be found in many parts. For example, most fresh water systems with similar conditions of temperature, salinity and water flow display identical habitats. Around the earth, all such habitats are grouped to form colonies, called the Biomes. In these regions, the relation of biotic and abiotic factors is more or less similar.

The plants, trees, shrubs, morphology, distribution in the area, climatic conditions, and geographical terrains are all identical. It differs from ecosystems as a biome is identified by means of the morphology and geographical distribution while the former relates to the genes, taxonomy and evolution.

Biodiversity is essential for understanding an ecosystem. A variation in the species in this region is accounted to biodiversity. Climatic condition is the key parameter that dictates the number of species in a region. Tropical and equatorial regions have greater genetic pools than the tundra. It describes the health of the system. The higher the diversity the greater is the resistance to changes within the region. Biodiversity finds application in many fields. It supports the quality of the air, temperature, climate, rain, sunlight and most significantly the nitrogen and carbon dioxide levels. When all these are maintained in the right proportions, there is a rapid rise in agriculture and forestry of the region. Apart from these features healthy soil, good health, and aesthetic value are all the various advantages.

A classic example of an ecosystem is a pond. In a pond ecosystem, the abiotic factors are the water flow, temperature and salinity. These are considered primary because depending on these factors organisms sustain in the region. In the biotic realm the most prominent are the green algae, as they manufacture the food and start the energy cycle. Predators in these areas are largely fish, flies, water snakes, frogs and tadpoles.

The energy is then circulated to the microorganisms. They are a great help to the environment as they help in the purification of the water, nutrient cycle and provide habitat. The pond habitat is disrupted with the addition of chemicals, an increase in temperature and a change in the number of species. Thus, a slight fluctuation can change the balance in the ecosystem.

Success of Ecosystems

Ecosystems predominantly result in the production of biomass, which is a key factor for the energy cycle. The photosynthetic process requires water molecules and sunlight for this production. So if the climatic conditions change then biomass is affected adversely. Thus annual precipitation and sunlight have to be consistent. Natural ecosystems and biomass production depend largely on nitrogen availability.

This can happen only if the litter from the plants is effectively recycled by the microbial. Decomposition releases the nitrogen into the atmosphere and soil, which is again utilized by plants during photosynthesis. Thus, the relation between the green, predators and the microbial is exponential.

Now when an ecosystem plays such a vital role, any disturbance from the original adversely effects the entire cycle. For example, if the carbon dioxide levels deplete due to migration of consumers from the region, the photosynthesis is depleted. The reverse is the scenario when trees and plants are felled.

The population in the region depends largely on conditions like food, reproductive possibilities and dispersal. Ecosystem shift is gradual as the diversity on earth is vast. Though sub-populations perish the balance is maintained, as speciation is another common factor in ecosystems. This maintains the equilibrium.

Importance and Restoring of Ecosystems

The essence of ecosystems is to understand the living and nonliving forms of the region. This has helped in maintaining a sustainable production of fuel, fiber, agriculture, biochemistry of the energy cycle and others. Tourism, global warming and eradication of soil erosion causes have all been substantially enhanced through ecosystem study. Maintaining a healthy ecosystem has been mediated through a number of services. Water filtration, forestry, agriculture, prevention of soil erosion, maintaining strong nitrogen and oxygen cycles are all services that are carried out to assist ecosystem development.

Ecosystems follow certain characteristics when attempts to restore them are carried out. These not only include vast geographical regions but also small gardens, parks, inland pools, fish tanks and so on. Some are even microscopic. Preserving an ecosystem is an effort that can be started from such small biological communities. Little plants at the household also increase the effects greatly. Biotopes in an ecosystem work collectively.

They are not individual entities, so connecting them through artificial systems is a good practice. Restoration is done in large scales by reputed organizations. Restoration of ponds, lakes, private gardens and plantings along highways are all preservation methods.

This complete approach successfully flourishes the ecosystem. The success includes biological efforts, agriculture, traffic planning, engineering and many more. A healthy ecosystem with prosperous biodiversity is the root for a healthy life.

Conclusion

Plants, Animals, Microbial, Genes and evolution: all are fascinating to understand. Thousands of organisms are still researched and every day many more come to light. Apart from this biology is a field of exceptions. This makes scientific research a continual process.

All disciplines in Biology are interrelated. The attempt to understand the body structure, its functioning and correlation with other organisms is primarily directed to one source: Leading a healthy and happier life. This is the crux of all scientific methods, as any imbalance caused through the living and the nonliving associated with humans can drastically alter existence.

www.ingramcontent.com/pod-product-compliance
Lightning Source LLC
Chambersburg PA
CBHW071726170526
45165CB00005B/2178